GERMAN TANKS IN WORLD WAR I
THE A7V AND EARLY TANK DEVELOPMENT

Wolfgang Schneider
&
Rainer Strasheim

Schiffer Publishing Ltd

4880 Lower Valley Road • Atglen, PA 19310

SOURCES

Photographic material was kindly provided by the following:

— Australian War Memorial, Canberra
— Bavarian State Archives, Munich
— Daimler-Benz Archives, Stuttgart
— Jim Hall
— May Hundleby
— Imperial War Museum, London
— Theodor Larsen
— Musée Royal de L'Armee, Brussels
— National Archives, Washington
— Panzer Museum, Munster

— J.C.M. Hall
— Queensland Museum, Brisbane
— RAC Tank Museum, Bovington
— Anneliese Schmidt, née Volckheim
— US Army Military Institute, Pennsylvania
— Joseph Vollmer Legacy
— German Museum, Munich
— Military History Museum, Rastatt
— Schneider and Strasheim collections

Translated from the German by Dr. Edward Force,
Central Connecticut State University.

Printed in China
ISBN: 978-0-88740-237-1

This book originally published under the title,
Deutsche Kampfwagen im 1. Weltkrieg,
by Podzun-Pallas Verlag, 6360 Friedberg 3 (Dorheim),
© 1988. ISBN: 3-7909-0337-X.

We are interested in hearing from authors with book ideas on German military history.

FOREWORD

Publications concerning German tanks usually dwell on the types used by the Wehrmacht in World War II, the Panther and Tiger, as well as Panzer III and IV, or they offer a present-day technical evaluation of the Leopard 1 or Leopard 2. The activities of German tanks in World War I and the technical beginnings of early German tanks are touched on only marginally.

The "A7V" is always cited as an example in this context, although only twenty of them were produced. But they saw front-line service numerous times and founded, to some extent, the tradition of the Panzer troops, which has lasted for more than seventy years.

The captured British tanks (Mark IV and Whippet) that saw German service in considerably larger numbers are obviously ranked well behind them, and of course they did not prove themselves in action. This the ideal significance of the A7V can, in the end, be ranked more highly than its actual role on the battlefield.

It is unfortunate that the operational significance of the tank was recognized too late in Germany. Technically, it would have been possible to build tanks even before the war. The first link track patent (Holt-Caterpillar) from the USA dates from 1888. As in other countries too, the military use of chain-driven armored vehicles was studied in the German Empire and in Austria, but it was limited from the start to armored wheeled vehicles incapable of cross-country travel.

Nor did the work of independent inventors such as Burstyn awaken any interest. But during the war too, factual incompetence and the petty jealousies of government agencies and individuals contributed to a lamentable confusion of projects and designs. And of course the lack of experience with the new materials resulted in many mistakes and misconceptions.

For example, a lot of time was wasted needlessly by equipping already existing vehicles and tractors with (too-heavy) armored superstructures (such as the Orionwagen, Lanz cross-country vehicle, Benz-Bräuer motorized limber, Büssing caterpillar tractor and Steil-Schreitkufenwagen), instead of concentrating on new designs from the start.

It would shatter the boundaries of this study to treat these many failed projects in depth.

Thus the work at hand will deal chiefly with the A7V as well as the technically interesting or progressive types.

The authors would also like to express the hope that historical interest in this early phase of German tank development will be stimulated.

Long before the tank saw service on the battlefield, armored wheeled vehicles were used as weapons carriers or armored road vehicles. Their decisive cross-country use had not yet occurred to military theoreticians.

Tractors with tracks also existed before the turn of the century. Thus it happened that the first suggestions for the creation of off-road tanks were widely rejected by most military leaders, attached as they were to the traditions of infantry and cavalry. Several dozen different technical suggestions got nowhere.

In the German-speaking area too, a technical project of particular interest became known. It was that of Austrian Oberleutnant Gunther Burstyn, who suggested a "motor-vehicle gun" with a rotating turret. After he had presented it to the Austrian War Ministry in October of 1911 and aroused no interest, he applied for a German patent on it in Berlin on February 28,

1912. Its climbing and trench-crossing capability was supposed to be increased by pairs of booms at the front and rear, held in place by springs. The low weight of five tons for the vehicle, which measured 3.5 meters (without the booms), was remarkable.

Surely Burstyn cannot be called the inventor of the tank, as was later done euphorically, and in view of its technical practicality too there would have been plenty of difficulties, but this project doubtless incorporated ideas that pointed the way into the future and indicated a number of details that were only realized much later (such as the rotating turret).

Drawings from patent document No. 252,815 of Austrian Oberleutnant Gunther Burstyn.

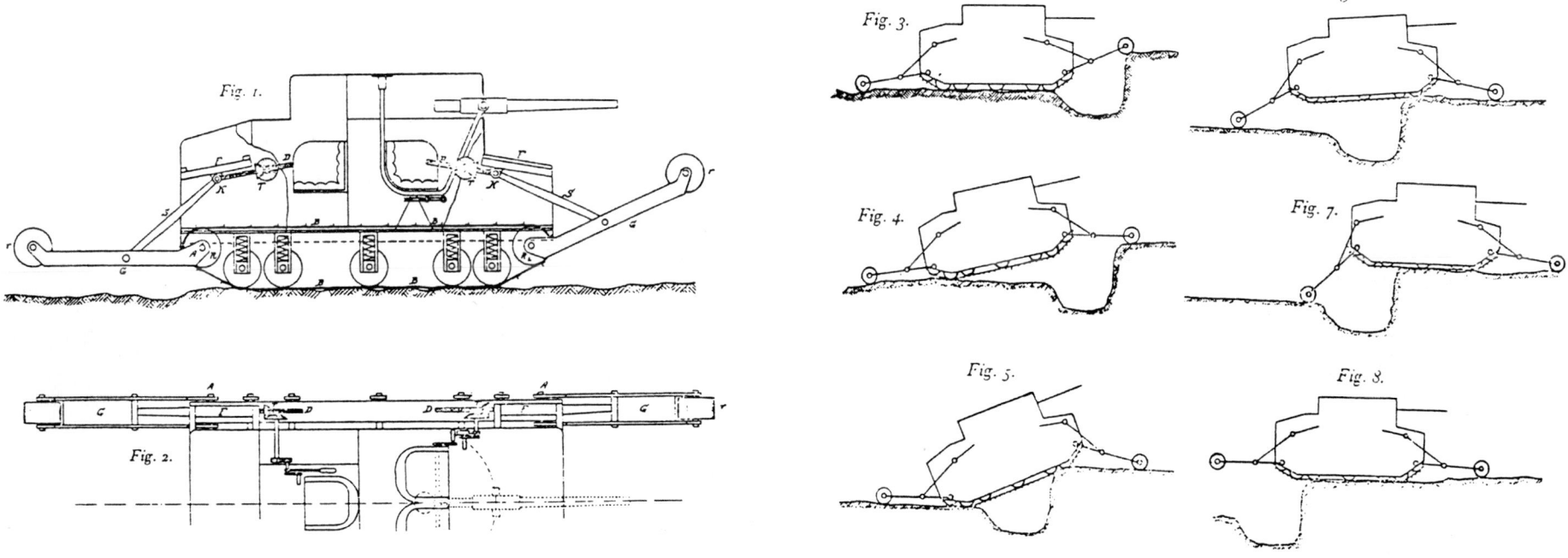

Fig. 1.

Fig. 2.

Fig. 3.

Fig. 4.

Fig. 5.

Fig. 6.

Fig. 7.

Fig. 8.

Above: A Marienwagen I chassis was fitted with an armored body. The overburdened vehicle could scarcely be steered, since the forward pair of tracks received no power.

Below: This armored vehicle on a Marienwagen II chassis, made by the Ehrhardt firm, had better driving characteristics, thanks to its normal front wheels.

Right: This rare contemporary photo document shows a Marienwagen II in action on Wilhelmstrasse during the Berlin uprising in January of 1919, being used by troops loyal to the government.

BREMER-WAGEN (MARIENWAGEN)

At the very beginning of the war consideration was given in Berlin to the construction of overland vehicles for use in transporting supplies in areas without roads. A contract was issued by the War Ministry to engineer Hugo G. Bremer on July 19, 1915 for the production of a so-called "overland wagon."

On October 6, 1916 a model was introduced in Neheim. Two pairs of tracks, of which only the rear ones were driven, were installed under a normal 4-ton truck. Fifteen of these "Bremer Wagons" were to be built in Marienfelde, near Berlin. But they were in no way satisfactory and were either rebuilt into normal trucks or developed further into the "Marienwagen."

In the Marienwagen I, the running gear of the tracks was supported by two springs from the frame, with the tracks running over small road wheels as well as two return wheels. Since the front tracks were, as before, not powered, and tended to slide off in turns, they were replaced by a front axle with normal wheels.

To meet the need for tanks, which had meanwhile become acute, the Bremer-Wagen was equipped with a body of 9-millimeter steel armor. But the truck was not capable of carrying this burden, and so the remaining chassis and the modified Marienwagen II were used as carriers for anti-aircraft and antitank guns. A fully tracked vehicle (Bremer-Wagen III) was designed later but not built.

DUR-WAGEN

In mid-1916 the Dürkopp Works in Bielefeld received a contract from the War Ministry to design an armored cross-country vehicle. The Dür-Wagen, as it was called, was equipped with two caterpillar treads, each driven by an 80-HP motor (without a differential). But this car also proved to be too weak to carry an armored body. Thus the two examples that had been built were used as transporters.

TREFFAS-WAGEN

At the request of the War Ministry, the Hansa-Lloyd Works in Bremen also tried to design a battlewagon. The Treffas-Wagen was finished on February 1, 1917. Thorough-going test drives showed it to be unsatisfactory. Meanwhile the decision had been made in favor of the A7V tank, and the only Treffas-Wagen was junked after one last demonstration in October of 1917.

Above: The Dürwagen without a body. It proved to be too weak for an armored body.

Above: Side view of the Orionwagen — fitted here with a mockup representing an armored body. It used a belt running on skids and a steered front wheel. It proved to be unsuitable for cross-country use.

Above: The Treffaswagen was an unusual vehicle, which moved on barrel-shaped hollow wheels. The small wheels at the rear were for steering.

5

Left: Testing the first A7V chassis, Berlin-Marienfelde, April-May 1917. The prototype was originally designed with two drivers (one for forward, one for reverse). This was later changed. The running gear, still without guiding panels to keep the tracks from getting untracked, can be seen. The main chassis frame extended well beyond the tracks fore and aft. The parallel upper and lower paths of the treads were the cause of the A7V's "low nose." In the prototype, only two coil springs were provided for the middle running gear (in production model 4). The links that connected the running gear are clearly visible.

Below: Introduction of the first chassis on April 30, 1917. The chief designer, Reserve Hauptmann and Oberingeneur Joseph Vollmer (in uniform), explains the vehicle to the Chairman of the VPK, General Friedrich (with goggles). Although the general was not overly pleased with the vehicle's performance, he agreed to further construction. The left track is being worked on here after a breakage. Note the crown-wheel housing projecting over the frame at the rear.

Above: The wooden body displayed by Daimler at Marienfelde on January 16, 1917 is shown here — in May of 1917 — already mounted on the prototype chassis. Note the grid on the nose. The later arrangement of the doors, hatches and windows is already to be seen here.

Above: Demonstration to officials of the OHL at the Mainzer Sand training camp on May 14, 1917. The vehicle now has a lengthened front grid and is waited to represent the later weight of the armor. It is noteworthy that the weight had been divided so evenly that the error of the "low nose" was not noticed.

Below: The Kaiser (third from left) views the prototype with its wooden body at the Daimler works in Marienfelde. Oberleutnant von Eichstedt, Pz.Kraftwagen MG.Abt.1, conducts the display on June 19, 1917.

The main frame of the prototype in the Daimler assembly hall at Marienfelde. The frame has not yet been reinforced but the vehicle has been converted to a one-driver system. Motors, fuel tanks and driveshafts have been installed, the exhaust systems are in place, likewise the guiding wheel with the track tightening apparatus. In the foreground one can see the steering wheel and the lever for the driver's compartment. The 4-ton trucks in the background were intended to be modified into Marienwagen (halftrack vehicles).

The driver's compartment: At right is the commander's seat, at left that of the driver, surrounded by the controls. The wheel for changing the engine speed (for wide curves), two clutch pedals, one speed control selector (for 3, 6 and 12 kph), at the same level two brake levers (one for each track), and at the far left two levers for forward or reverse drive (one per track). The handwheel at the far left behind the commander's seat is the starter.

THE A7V ASSAULT TANK

On November 13, 1916 the War Ministry gave the Motor Vehicle Technical Testing Commission (VPK) a contract to develop a tank (the term "Sturmpanzerwagen" was applied to it only on September 22, 1918). As a disguised name, the designation of the responsible department of the War Ministry, "A7V", was chosen.

With a gross weight of about thirty tons, it was supposed to be capable of running cross-country, spanning ditches up to 1.5 meters wide, and reaching a road speed of 12 kph. Its armament was to include one cannon each to the front and rear, plus several machine guns flanking them. A motor producing 80 to 100 horsepower was thought to be sufficient!

For its development, a technical commission was chosen, composed of military and business officials, directed by the VPK and Oberingenieur Vollmer.

The running gear design was made with the help of the Holt-Caterpillar Company in Bucharest.

On December 22, 1916 the War Ministry made the design drawings of the A7V tank (Vollmer type) public. Now two motors, each producing 100 HP, were included. The suggestion to give the A7V the highest priority rating for the production of war materials was rejected by the Army High Command (OHL).

Unfortunately, it was urged that the A7V should be used as a so-called "overland wagon" and only ten examples should be armored. At times the unrealistic demand was also made that the vehicle should be able to withstand direct hits from artillery shells.

The normal design, though, was retained except for the dimensions being increased (the length became 7.35 meters instead of 6.26).

The first presentation — with a wooden body — took place on April 30, 1917 at Marienfelde; it went off satisfactorily.

The gearbox made by Adler of Frankfurt am Main. The brake drums are set on the two driveshafts that extend backward. The large cogwheels that transmitted the power to the drive wheels projected up and down from the main frame and limited the A7V's ground clearance to only 20 cm.

The gearbox installed, here in Tank 506, "Mephisto." The covering plate of the rear fighting compartment is missing. While both brake drums of the prototype were synchronized by one activating lever, the production model had one lever for each brake drum, seen here to the right and left of the gearbox (with thickened sections).

The constant changes that were made during the construction of the A7V were intensified by the question of its armament, and it took a long time until a suitable cannon could be found. All of this lasted until the spring of 1918.

Finally the Belgian 5.7-cm casemate cannon, which had a slight barrel recoil. Instead of two cannons, just one, pointing forward, was finally decided on, plus six machine guns instead of four.

As early as July of 1917 the first A7V tanks were supposed to be ready for service. On May 14, 1917 the OHL's decisive presentation of an A7V with ten tons of ballast (and wooden tank body) took place in Mainz.

Raw material shortage and delivery delays held up series production, and technical defects in the drive train, running gear and armor plate also had to be cured.

The first A7V was finished only at the end of October 1917. On September 20, 1917 the War Ministry ordered the formation of Assault Tank Units 1 and 2. Each was to consist of five officers and 109 non-commissioned officers and enlisted men, plus five A7V tanks and nine wheeled vehicles. The formation of Unit 3 was ordered on November 6, 1918. Unit 1, reported ready to march in Berlin on January 5, 1918, was transferred to a driving school at Sedan. All in all, though, the training level of the quickly assembled crews was only moderate, and cross-country use soon made the tank's technical shortcomings (too-low bow, insufficient ground clearance, insufficient engine power, running-gear problems) show up all too often.

The decision to build 38 A7V's, which had meanwhile been made, was rejected and the number set at twenty, in order to await combat experience. Thus the end of the tank was foreseen even before its first action!

After more thorough crew training, the tank's baptism of fire took place near St. Quentin on March 21, 1918 (see chapter "Combat History"). All in all, the A7V proved itself without problems in the fairly passable terrain and was even superior in some ways to the British tank types then in use; the main problem from the beginning was the poor durability of the running gear. The tanks were — if at all — only ready for a few hours of action and soon needed a general overhaul. At this time the so-called Bavarian Army Vehicle Park 20 at Charleroi had been set up, where captured British tanks were also prepared for service in the field.

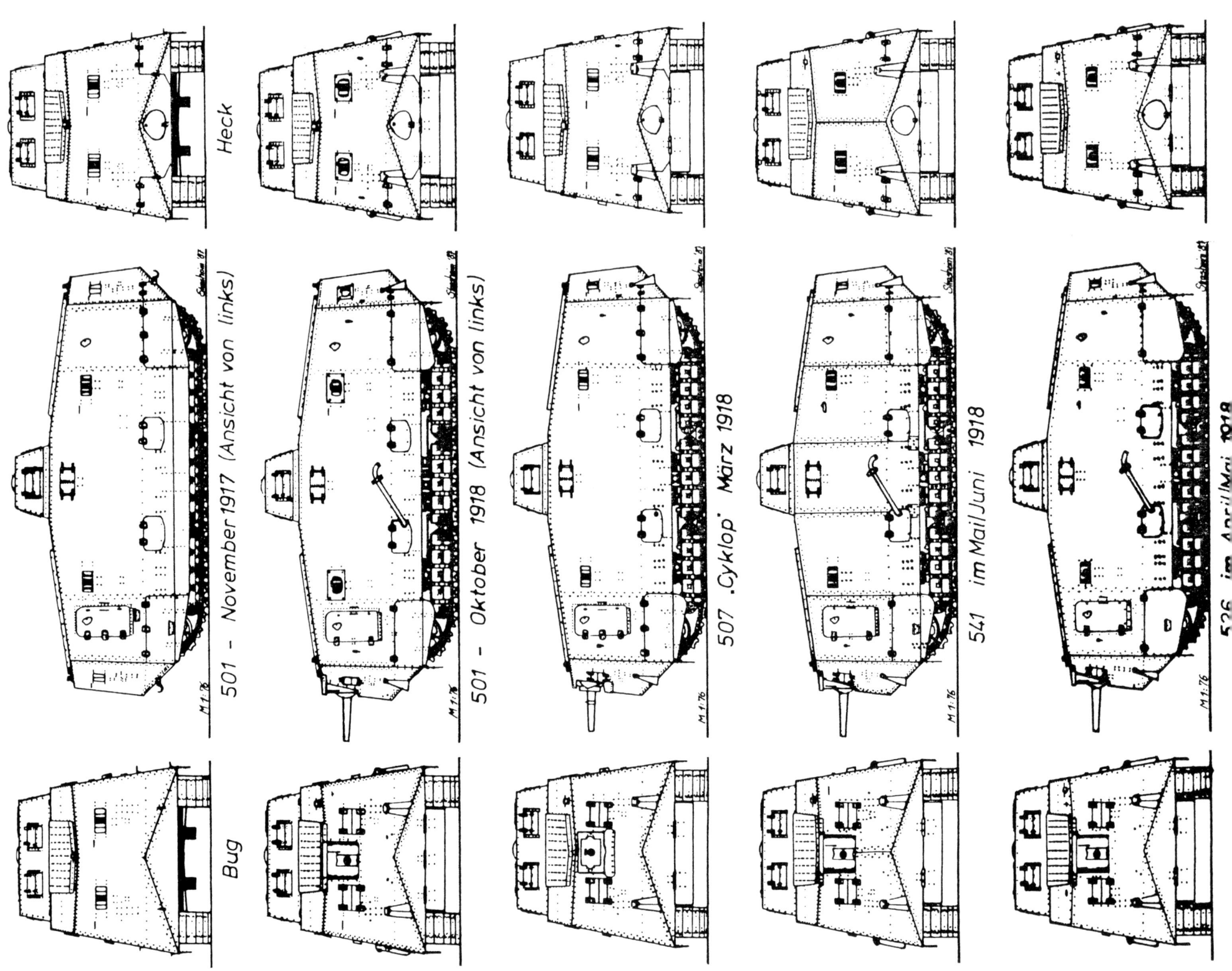

Heck

Bug

501 – November 1917 (Ansicht von links)

501 – Oktober 1918 (Ansicht von links)

507 „Cyklop" März 1918

541 im Mai/Juni 1918

526 im April/Mai 1918

M 1:76

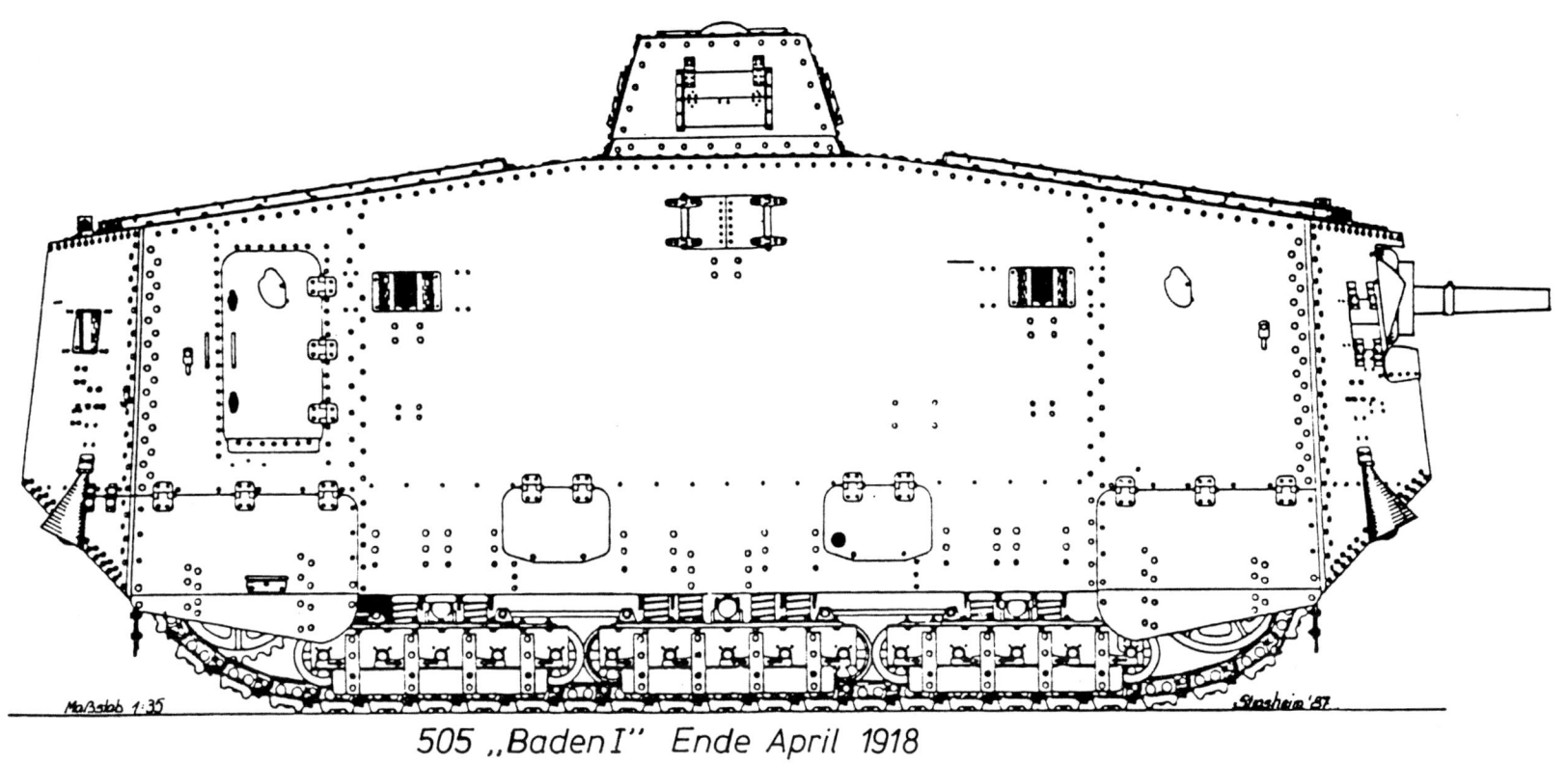

505 „Baden I" Ende April 1918

Scale 1/35505 "Baden I" end of April 1918

About the ten construction batches of the A7V: Originally ten sets of armor were ordered, of which Krupp and Röchling (Dillingen foundry) each supplied five. The armor plates from Krupp were delayed, though, and had to be cut to fit. This resulted in A7V tanks with five-piece side and two-piece bow and stern armor. In the first batch (Röchling armor) were Tanks 501, 502, 505, 506 and 507. The first to be finished was 501 — though without a cannon — and it showed a few special features. 502, 505, 506 and 507 were equipped with the 5.7-cm cannon on a trestle mantelet. The chassis of 502 became unusable in March of 1918 — presumably after the frame broke. Its armor was promptly fitted to Tank 503, which Unit 1 had used until then as a recovery vehicle. The first batch with Krupp armor included Tanks 540, 541, 542, 543 and 544. The chassis of 544 soon became unusable too, and the armor for it was transferred to the second recovery tank, 504. The tanks with Krupp armor were already equipped with the socket mantelet (originally developed for the A7V-U) for the gun. The tanks of the second production batch included numbers 525, 526, 527, 528, 529, 560, 561, 562, 563 and 564. These also had socket mantelets, as well as fewer screws and rivets in their sheet-metal parts. All chassis and armored bodies were handmade and thus show considerable differences in their measurements. The drawings give an overview of the different variations.

Right: Tank 501 on a test run at the Daimler works. The first finished A7V tank had no cannon as yet. Its armament consisted of four MG 08/15 machine guns and two flamethrowers. This picture was taken early in November 1917. Bote that 501 still has no hole for the exhaust. The free-swinging panels at the bow and stern, which were built in later — to protect the unarmored underbody — are missing, as are the coverings for the towing hook and the guiding panels to prevent the tracks from coming off the road wheels.

THE A7V IN COMBAT

In January of 1918 Unit 1, the first German "Sturmpanzerkraftwagen" unit, rolled onto the field. But it took almost a quarter of a year until it was ready for combat.

On March 21, 1918 it saw action under the leadership of Hauptmann Greiff near St. Quentin. Of its five tanks, one went out of action prematurely and two shortly after the attack began. The other two, 501 and 506, took an effective part in the battle. The small number of vehicles that saw action, along with the high percentage of breakdowns, prevented any definite evaluation.

During March of 1918 Units 2 and 3 also became ready for service.

A planned use of Units 1 and 2 on the Ailette by the 7th Army did not take place, as the French had voluntarily withdrawn across the brook.

Below: A view of the Daimler A7V production plant at Marienfelde, which had been built for this purpose, and where the Marienwagen was also built. In December of 1917 the tanks of the first production batch (Röchling armor) were almost finished. Tank 505 is shown just before completion. The hole for the exhaust is easy to see. The panels over the road wheels were later removed by the troops in order to improve the self-cleaning of the running gear.

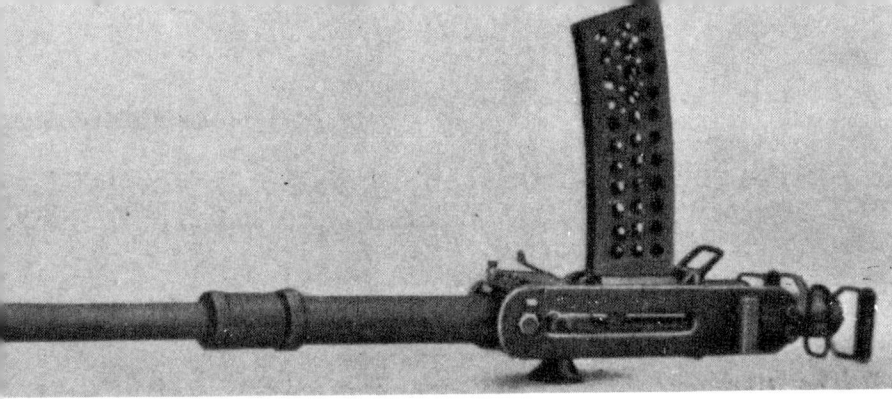

Above: The 2-cm Becker cannon originally intended to be used proved to be too unreliable and too ineffectual on its target. The plan called for two of these weapons and four 08 machine guns for an A7V tank, in which cannons and machine guns were to be interchangeable.

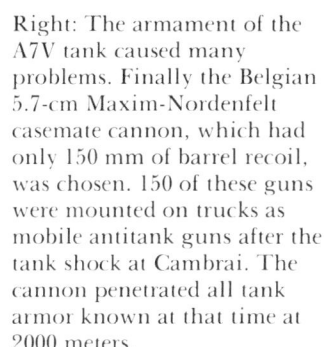

Right: The armament of the A7V tank caused many problems. Finally the Belgian 5.7-cm Maxim-Nordenfelt casemate cannon, which had only 150 mm of barrel recoil, was chosen. 150 of these guns were mounted on trucks as mobile antitank guns after the tank shock at Cambrai. The cannon penetrated all tank armor known at that time at 2000 meters.

Left: Tank 506 (later called "Mephisto") being repaired at Charleroi early in April 1918. The armored body has been lifted off, the cannon detached and moved backward. The balance weight has been attached. It was aimed by using fore- and backsights. The backsight was inside the body, the foresight on the barrel. The orientation of the gunner was thus very good, but the large opening in the side aiming panel was very sensitive to infantry weapons fire.

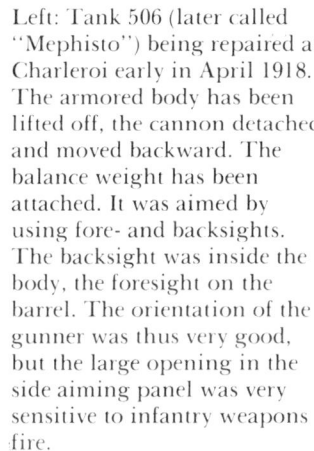

Right: The trestle mantelet developed by the Artillery Testing Commission for the 5.7-cm Maxim-Nordenfelt gun. The left handwheel is for traverse aiming, the pierced right one for elevation. This picture was taken of "Mephisto" in the Queensland Museum, Australia. The gun and balance weights are missing.

The socket mantelet for the 5.7-cm Maxim-Nordenfelt, actually developed for the A7V-U. It was soon learned that this mantelet was also suitable for the A7V tank and for captured tanks as well. It was built at the artillery workshop at Spandau. Since a suitable means of aiming had to be developed separately, the trestle mantelet was improvised for the first production batch (Röchling armor); later (mid-1918) it too was replaced by the socket mantelet.

Below: The gun mounted on the socket mantelet. Over the recoil brakes is the direction indicator with which the commander could give the crew the general direction to the target. At right above the port is the indicator light, also used by the commander. White meant attention, red: fire, no light: load the gun.

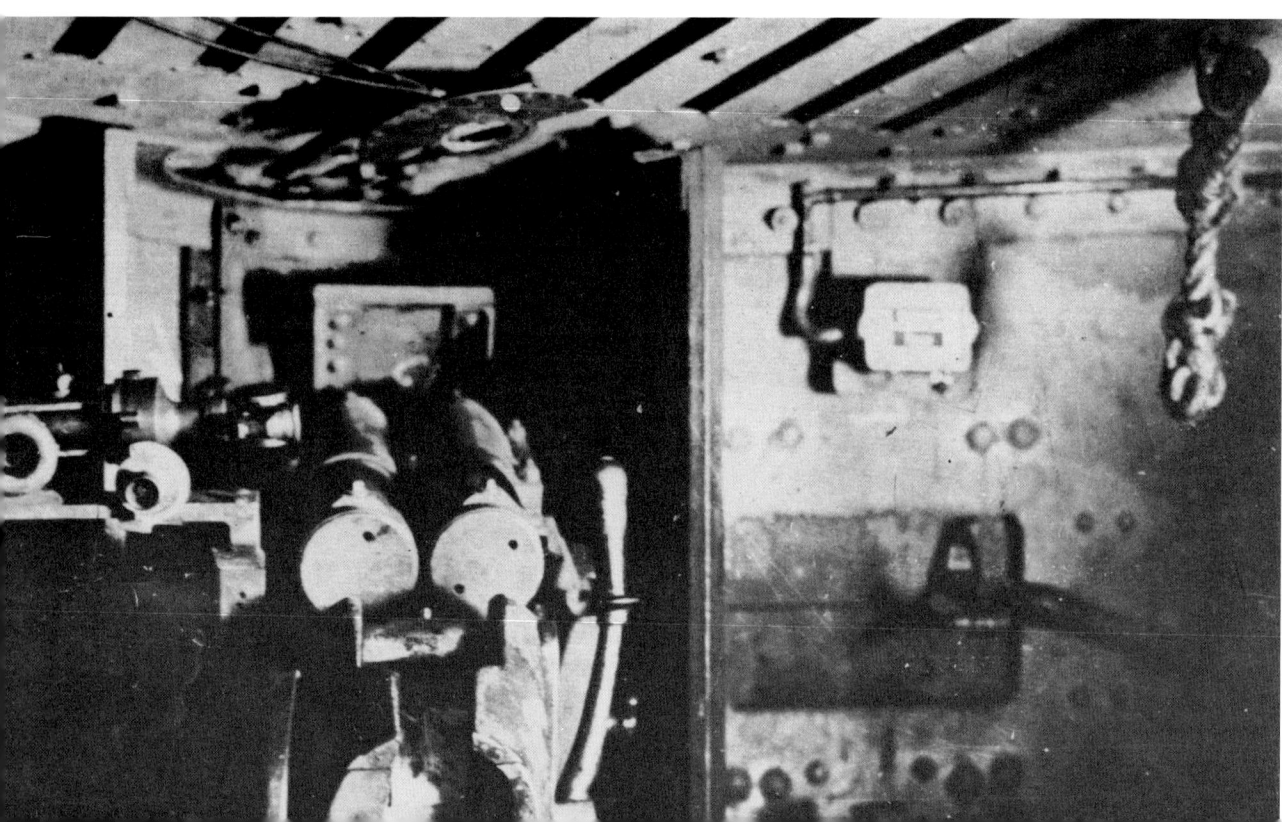

Above: The pierced handwheel was used for elevation, the other for traverse aiming; over them is the scope. In overland travel, the gunner could easily lose his orientation on account of the tank's movements. The recoil brakes are attached above the barrel, and the firing lever is at right.

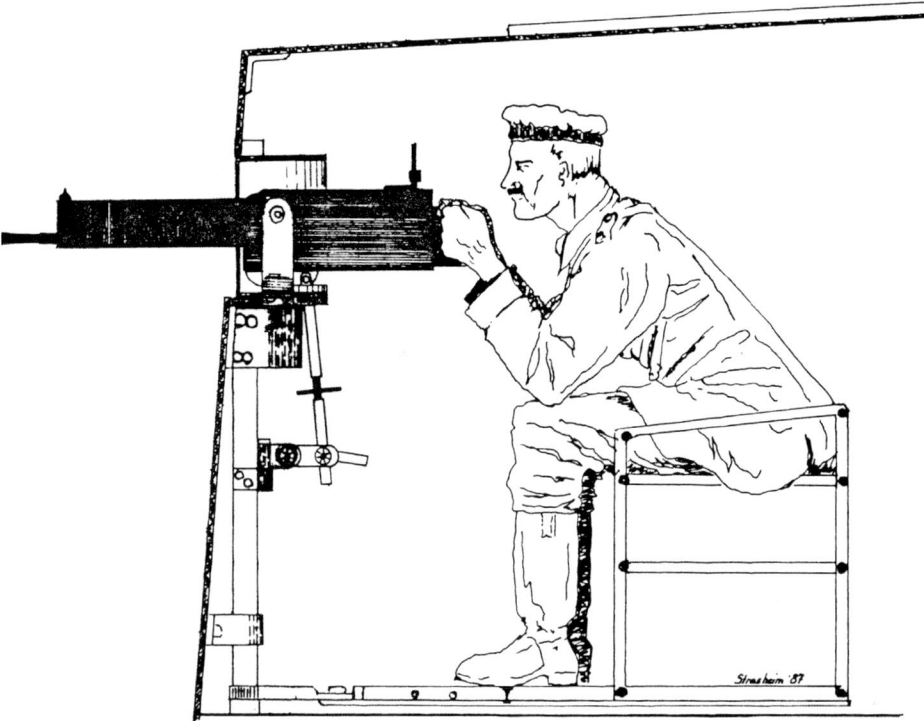

A7V – Maschinengewehrstand

Above: A British soldier demonstrates the operation of a stern machine gun in Tank 542, "Elfriede." Over the gun is what remains of the indicator light box. The chalked "Jung" is the name of the gunner assigned to this gun in the battle of Villers-Bretonneux. The cartridge-belt cases were stacked in the machine gunner's seat. At upper right is the schematic drawing of a machine-gun position. The gunner in his seat was firmly attached to the gun and turned with its mantelet. Forty to sixty type 15 cartridge-belt cases, each holding a belt with 250 cartridges — a total of 10,000 to 15,000 rounds — were carried. Three to four hundred shells for the 5.7-cm gun were carried in battle. (The equipment list included only 180 shells.) Of them, about 50% were canisters, 30% antitank shells and 20% explosive shells with impact fuses.

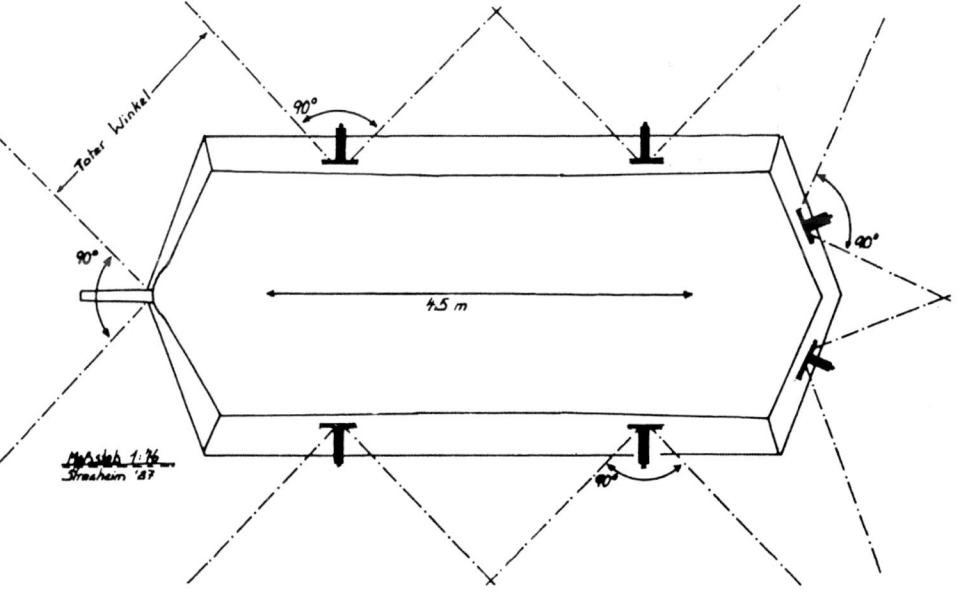

Right: The fields covered by the weapons of the A7V. All weapons could be aimed so that they covered the terrain as of about 4.5 meters from the tank. The dead angles at the bow had to be covered by zigzag driving. For its time, the A7V possessed a murderous firepower. For shock-troop assignments it also carried a light 08/15 machine gun, several rifles and hand grenades. Flamethrowers, although included in the planned equipment, were not used.

15

Above: Linen overalls including asbestos fibers were issued to the A7V crews. The caps seen in the picture were not worn in combat. The personnel coming from the motorized troops (drivers, mechanics) often wore leather helmets, such as were worn by motorcycle couriers and aviators.

Left: Three motorized soldiers in front of Tank 560. While machine gunners and artillerymen often continued to wear the uniforms of their old regiments, the drivers wore the insignia of their original troop unit, the guard Motor Vehicle Battalion. The man at right is a non-commissioned officer.

Above: A motorized soldier of Unit 1 poses before Tank 560. The leather helmet is the second type used. The shrapnel-protection mask was either captured or copied from the British. It was scarcely ever used except for photographs.

Lower left: Tank 507 passed through a suburb of St. Quentin to an assembly point after the battle. For camouflage, the tanks were painted in irregular spots and stripes of red-brown, light green and lime yellow.

Left: The first A7V action took place on March 21, 1918 — as part of the "Michael" offensive — south of St. Quentin. Tank 506 —photographed from 501 — rolls forward. Note the Iron Cross emblem on the side.

Below: Tanks 501 and 506 went to Charleroi for overhauling as of March 21, 1918. This picture was taken there at the Bavarian Army's Vehicle Park 20. Unit 1 used the death's-head emblem on the bow of the tank. All the preparations have been made to lift off the armored body of 506. Tanks 505 and 507 went back to Beuville after 3/21/1918 to be used in training Units 2 and 3.

Left: A tank of Unit 1 — probably 526 — in St. Gobain. The crew has gotten out the kitchen utensils and is waiting for the field kitchen.

Below: Since they did not go into action, Units 1 and 2 were taken back to Charleroi by train. Here Tank 560 (Leutnant Volckheim) is seen on a makeshift ramp. Such a ramp could be built by a A7V crew in about three hours, using its own materials. 560 is completely painted field gray.

On April 24, 1918 all three units, with the 2nd Army, saw action at the country village of Villers-Bretonneux, a forest, the Bois d'Aquenne, southwest of it, and the village of Cachy, which were occupied by British troops.

Tank 540 of Unit 2 broke down even before being loaded on the train, Tank 503 of Unit 3 broke down with a cracked cylinder head during preparation. The remaining thirteen tanks were used in three tank groups:
Group 1 (3 tanks of Unit 1), under the command of Oberleutnant Skopnik, with the 228th Infantry Division against Villers-Bretonneux.
Group 2 (4 tanks of Unit 3 and two of Unit 1), under the command of Oberleutnant Uihlein, against the southern edge of Villers-Bretonneux and the Bois d'Aquenne.
Group 3 (4 tanks of Unit 2), under the command of Oberleutnant Steinhardt, with the 77th Reserve Division against Cachy.

Moving forward in the thick morning fog, the tank groups at first operated very successfully, as the poor visibility hindered the action of the British field artillery.
Group 1 (Tanks 526, 527 and 560) helped the 228th Infantry Division to take Villers-Bretonneux by midday.
Group 2 had been divided into two troops to support the 4th Guard Infantry Division.
Troop 1 (Tanks 505, 506 and 507) advanced against the southern edge of Villers-Bretonneux. Tank 506 stopped with plugged jets; when it was running again, it tipped into a large shell crater.

Left: Tank 527, "Lotti", (Leutnant Vietze), rolls through Villers-Bretonneux, which was captured around midday with the effective support of Tank Group 1. The British infantry showed that it could not equal the German tank attack, and the artillery was handicapped by the heavy fog and could only intervene after Villers-Bretonneux had been taken.

Below: The crew of 560 at the assembly point after the battle. The commander, Leutnant Ernst Volckheim, is fifth from left. The regiments of the 228th Infantry Division spoke very positively about the support given them by Tank Group 1. The infantry's losses remained encouragingly low, since the armored monsters affected the British troops' fighting spirit negatively.

Right: Tank 505, "Baden I", (Leutnant Hennecke) of Unit 3 (Tank Group 2) at the south side of Villers-Bretonneux. Despite ammunition defects, the tank remained in service, with an machine gun pushed through the right front port. Tanks 505 and 507 (Leutnant Bürmann) worked perfectly along with Reserve Infantry Regiment 93 of the 4th Guard Infantry Division. All attack targets were taken according to schedule.

Lower right: Tank 507, "Cyclops", in its way back to Charleroi, early May 1918. Note the black-white-red flag at the rear. The equipment of an A7V unit included, besides five tanks, nine trucks, two cars, one field kitchen trailer and one motorcycle.

Troop 2 (Tanks 541 and 562 of Unit 1 and 501) was to move against the Bois d'Aquenne. stopped before the first British trenches with an overheated motor and finally moved backward.

Tank 562 had suffered seized brakes and gearbox damage when its driver was wounded, and part of its crew was lost in a shock-troop action. After it was made mobile again, 562 advanced toward the Bois d'Aquenne. At first 541 broke the resistance in a heavily fortified farmyard south of Villers-Bretonneux and then joined 505 and 507, with which it watched over the advance of the infantry into the Bois d'Aquenne.

20

Upper left: Tank 561, "Nixe" (Leutnant Biltz), before April 24, 1918. When it was loaded for action at Villers-Bretonneux, Unit 2 was just removing the panels over the running gear. In action, Leutnant Biltz took a wrong turn at first, then fought in the first tank battle in history. After it, Tank 561 could be recovered, but it remained unusable and had to remain out of service, since no spare parts for the A7V were available.

Above: Tank 543, "Hagen", broke down before the battle and later joined Unit 3, where it was renamed "Adalbert." The picture shows it with Unit 3. For the battle at Villers-Bretonneux, all the tanks of the first production batch and six of the second were delivered to the troops.

Left: Tank 525, "Siegfried" (Leutnant Bitter), halted the advance of British Whippet tanks east of Cachy (along with an escort battery of the 4th Guard Infantry Division). Tank 504, "Schnuck", was also there but did not take part in the tank battle. Here we see "Siegfried" and its crew after the battle.

21

Upper left and right: Tank 542, "Elfriede" (Leutnant Stein) of Unit 2 (Tank Group 3) tipped into a shell crater on April 24, 1918. Leutnant Stein was killed defending the tank. 542 remained in no-man's-land until May 15, 1918. Then it was reached by soldiers of the 37th (Moroccan) Division and recovered with the help of A Company of the British 1st Tank Battalion. The pictures on this page show the tank lying on its side in the sand. The elements of the undercarriage are clearly visible: The carriers of the main frame, the rod-shaped leading arms and the underslung transverse members, the last recognizable by their triangular panels. At the bow is the armor of the gun area and fuel tank, in the middle the two exhaust mufflers, at the rear the gearbox. One of the two large crown-wheel housings has a metallic shine; it has been rubbed smooth because of its closeness (only 20 cm) to the ground. French and Moroccan troops pose with the captured tank, which they covered with chalk inscriptions. Before the bow, the houses of Villers-Bretonneux can be seen in the background.

Left: "Elfriede" — seen here from the bow and stern — could be made mobile again with little trouble after she had been towed behind the Allied lines. The Allied test reports say that the A7V moved as fast as a trotting horse over smooth ground but could neither cross a 2.40-meter ditch nor climb a 1.20-meter bank. The driver's and commander's views, they said, were remarkably limited; they could not see the ground for the first nine meters in front of the tank. All the openings and dents were very sensitive to the lead and steel splinters of striking infantry bullets. The side armor could be pierced at five meters by the French APX gun (similar to the German Smk).

Right: French driving tests with Tank 542. A view of the turret, with the driver at the right, the commander at the left.

Below: After the end of testing, 542 was displayed at the Place de la Concorde in Paris. Plates were cut out of the armor for shooting tests. This picture was taken in December of 1918.

Right: Since it had been blown up deliberately on the night of April 25, 1918, Tank 506, "Mephisto", was not recovered, though it still lay behind the German lines until June. On July 14, 1918 the Australians recovered it after they took the German positions. The tank was painted with splendid designs and then displayed behind the front. "Mephisto" had the figure of a red devil, carrying away a British Rhombus tank, on its bow plate as a mascot. As "revenge" the Australians painted a British lion, grasping 506, on its side.

The effect of the explosive charge, which was set off in the forward fighting compartment, can be seen clearly. At the left side of the picture, the front radiator and both fuel tanks, one almost undamaged, the other bent out of shape under the radiator, lie by the bow.

Group 3, while advancing, had to go around a heavily shot-up and ravined woodland.

Tank 542 went too far to the north and tipped over into a sandpit. Tank 561, which also went too far north, was moving toward Cachy in the fog when it suddenly saw three British tanks facing it. The two British machine-gun tanks were poorly equipped, but the cannon tank, which the Germans thought they had already destroyed, scored a hit with its 25th shot that penetrated the right front port and knocked out the gun crew. The commandant gave the order to abandon. 561 was hit two more times on the right flank.

Since the motors were still running, the crew boarded it again after some time — the British cannon tank had been knocked out by

a mine thrower. It was possible to bring the tank back about two kilometers before the engines finally seized. Tanks 525 and 504 reached their combat area east of Cachy as planned. The 77th Reserve Division, inexperienced on the western front, did not make progress, for an attack of seven British Whippet tanks confused them thoroughly. Tank 525 moved against the Whippets at once, and at the same time a shock battery of the 4th Guard Infantry Division opened flank fire. Four Whippets were left burning; three were able to flee. At night two Australian brigades made a counterattack. Tank 561 could still be recovered in time; 542 was to be blown up. Instead, the demolition command blew up Tank 506.

Thus the morning of April 25 found Tank

542 unharmed in no-man's-land and 506 behind the German lines, unusable.

Tank 542 was towed away by the French and British in May. The tank was still driveable and was tested thoroughly.

In July, after the German lines had slowly been pushed back, Tank 506 was captured by the Australians and finally taken to their homeland, where it can still be seen today.

Tank 542 was displayed in Paris until 1919 and finally junked. Tank 561 could not be repaired and was scrapped.

Unit 2 went into combat near Reims on May 31, 1918 and lost two tanks prematurely. In the attack the lead tank, No. 529, was shot up by the French artillery and lost. The two remaining tanks turned around.

Tank 529 was recovered by the Americans

"Mephisto" was sent to Australia in 1919. It is seen at left in front of the Queensland Museum in Brisbane in September of 1919. The tank sat there rusting and ignored for fifty years. In 1972 it was sandblasted and repainted. The right picture shows it in front of the Queensland Museum in 1983. For the Australian Bicentennial it was completed externally and painted colorfully.

in 1919 and taken to the USA, where it was scrapped in 1942.

On June 1, 1918 Unit 1 attacked northwest of the Fort de la Pompelle. Two tanks broke down before the combat began. Tank 527 got stuck and was destroyed by a direct artillery hit. Tank 526 also got stuck and was recovered in such a damaged condition that there was no question of its further use. Tank 560 broke off the attack.

Tank 527 remained where it was until 1921 and then was junked when the country was cleaned up. Tank 526 was scrapped.

What remained of Unit 1 and Unit 3 saw action with the 18th Army south of Noyon.

Unit 1 lost Tank 560 temporarily when it was hit by artillery fire. In the attack, 562 rolled into a big hole and could be recovered only several days later. Tank 541 gave up the attack after engine and gearbox damage. Unit 3 fulfilled its assignment satisfactorily. Only Tank 564 got stuck in a village and was put out of action.

When Units 1 and 2 saw action with the 7th Army on July 15, 1918, everything went off according to plan and without losses.

On August 31, 1918 Units 1 and 2 were to support a counterattack near Frémicourt. Unit 1 did not reach the combat area on time, and the attack of the division that it was to support was smashed. Unit 2 attacked without a chance of succeeding. Tank 504 was destroyed by German artillery, 526 got stuck while trying to retreat, 562 had broken down in advance after being hit by air bombs, 563 limped back to the German lines with mechanical damage.

Tanks 504 and 528 were captured by the New Zealanders. After the war they were displayed in Britain and then junked in 1919.

After the disaster at Frémicourt, Unit 2 was disbanded de facto by being subordinated permanently to Unit 1.

Unit 3 took part in a counterattack of the 3rd Army at St. Etienne on October 7, 1918. All the tanks were damaged during the battle.

The attack had to be broken off, since the bridges over the Arne had already been blown up.

Left: The two engines of Tank 506. The right one is numbered MD 9088 on its cylinder block and MD 9483 on its crankcase. The rear radiator can be seen between the cylinders; the exhaust pipes are in the center.

Lower left: The passageway at right next to the engine room, looking toward the stern. The height of the passage is barely 1.60 meters.

Below: The rear end with the gearbox housing and escape hatch. What remains of the machine-gun mantelet is below the gun port.

The last A7V action was taken by the strengthened Unit 1 near Iwuy on October 11, 1918.

Tank 562 had already broken down and been scrapped. The attack of the five remaining tanks, 525 (Lt. Wagner), 563 (Lt. Goldmann), 501 (VzFw. Lommen), 540 (Lt. Schück) and 560 (Lt. Volckheim), was thoroughly successful in eliminating a British breakthrough. A track of Tank 560 broke and the tank was blown up.

Tank 562 was captured by the British and scrapped in France.

Thus ends the combat history of the A7V. The few remaining tanks were sent back to Erbenheim, near Wiesbaden. There the units were quickly disbanded after November 11.

The tanks fell into the hands of the advancing French troops and were junked.

During the Berlin Revolution an A7V copy, the tank "Heidi", saw service in Berlin. It had to be turned over to the Entente that summer, and was likewise scrapped.

Left: After Villers-Bretonneux several tanks were sent to the armies on the western front for demonstrations. Here Tank 543, "Adalbert", is seen. Until May of 1918 a single Iron Cross was painted on the sides; then it was doubled, and the number of the tank in its unit was painted between.

Right: Tank 540, "Heiland", was not used by Unit 3. The tank was turned over to Unit 1 in the summer of 1918 after being repaired.

Above: A view of the command post of "Wotan." The hand accelerator is beside the steering wheel.

Upper left: Tank 529, "Nixe II" (Leutnant Biltz), was lost north of Reims on May 31, 1918.

Left: After Tank 529 was lost, Unit 2 had only four tanks left in June and July of 1918. From left to right: 563 "Wotan", 528 "Hagen", 525 "Siegfried" and 504 "Schnuck", on display in a French village. The unit leader, Oberleutnant Steinhardt, is wearing goggles.

Above: Tank 563, "Wotan" (Leutnant Goldmann), a typical tank of the second production batch. It was painted a solid field gray but covered with dust.

Upper left: Unit 1 in action near Fort de la Pompelle (east of Reims) on June 1, 1918. Tank 526 (Oberleutnant Skopnick) heads for its departure position; the assistant commander, Leutnant Philipp, walks ahead.

Above: Tank 527 (leutnant Bergemann) was lost on June 1. It got stuck and later took a direct artillery hit in its turret. Note the Iron Cross on the ventilator grille.

Left: The remaining tanks of Unit 1 were loaded on June 5 for service on the Matz. Here Tank 560 (Leutnant Volckheim) is seen with its assistant commander, Leutnant Bergemann.

Above: On June 9, 1918 Oberleutnant Skopnik and Leutnant Bartens of Unit 1 were killed by artillery fire. Tank 560 was knocked out by grenade launchers. Only Tanks 562 (shown here) and 541 attacked.

Above: On August 31, 1918 Unit 2 lost two tanks at Frémicourt. Under fire from German artillery, Tank 528 "Hagen" (Leutnant von Jamrowski) got stuck and was abandoned. "Hagen" fell into the hands of the New Zealanders, who recovered it and turned it over to the British.

Upper right: Tank 504 "Schnuck" (Leutnant Kunze) took two frontal hits from German artillery and was abandoned by its crew. Its subsequent fate was like that of "Hagen."

Right: "Hagen" in London (Horse Guards Parade) in 1919. The camouflage paint (red-brown and lime yellow) has largely flaked off.

As of September 1918 the A7V tanks were painted in bright colors (frost green, lime yellow and red-brown over field gray). Here is Tank 501 of Unit 1 in a drill.

Tank 501 has gotten stuck. At the same time as the introduction of camouflage paint, the Iron Crosses were replaced by plain crosses.

Below: Tank 501 drawn back into the brush during a drill. As of September 1918 the death's-head was painted on the bow of all A7V units; previously only Unit 1 had used it.

Below: Tank 501, side view. The tank is now equipped with a socket-mounted gun. Note the three hinges on the free-swinging panel.

Left: The tanks of Unit 3 were so worn out from use that Tank 503 finally had to be abandoned. To make it easier to recognize, the crosses were surrounded with white paint.

Right: Tank 503 was towed back to the Charleroi area, but BAKP 20 had already been sent back to Germany. But the tank could not be repaired and fell into the hands of the British, who scrapped it on the spot.

34

THE A7V TANKS, THEIR COMBAT AND THEIR FATE

Number	Name	Unit	Origin	Notes
501	Gretchen (teilw.)	Abt. 1, dann Abt. 3, dann Abt. 1	1. (Röchling)	With army till war ended
502/503		Abt. 1, dann Abt. 3	''	Abandoned 10/1918, junked
504/544	Schnuck	Abt. 2	1. (Krupp)	Taken by British, junked 1919
505	Baden I (teilw.)	Abt. 1, dann Abt. 3	1. (Röchling)	With army till war ended
506	Mephisto	Abt. 1, dann Abt. 3	''	Captured by Australian troops, in Queensland
507	Cyklop	Abt. 1, dann Abt. 3	''	With army till war ended
525	Siegfried	Abt. 2	2.	''
526		Abt. 1	2.	Stripped
527	Lotti (teilw.)	Abt. 1	2.	Lost at Reims, junked 1922
528	Hagen	Abt. 2	2.	Taken by British, junked 1919
529	Nixe II	Abt. 2	2.	Taken by Americans, junked 1942
540	Heiland	Abt. 3, dann Abt. 1	1. (Krupp)	With army till war ended
541		Abt. 1	1. (Krupp)	With army till war ended
542	Elfriede	Abt. 2	1. (Krupp)	Taken by French, junked 1919
543	Hagen Adalbert König Wilhelm	Abt. 2, dann Abt. 3	1. (Krupp)	With army till war ended
560	Alter Fritz (teilw.)	Abt. 1	2.	Blown up 10/11/1918
561	Nixe	Abt. 2	2.	Stripped
562	Herkules (teilw.)	Abt. 1, dann Abt. 2	2.	Taken by British, junked
563	Wotan	Abt. 2	2.	With army till war ended
564		Abt. 3	2.	With army till war ended

The last attack made by Units 1 and 2 took place on October 11, 1918. Tank 560 had to be blown up after breaking a track. Tank 562 "Herkules" (now in Unit 2) became unusable beforehand and was stripped for spare parts. It fell into British hands and was demolished.

Below: The end: an A7V tank is scrapped.

The A7V tanks of Units 1 and 3, some ten tanks in all, were transferred to Erbenheim, near Wiesbaden, in November of 1918, and scrapped there in December by the French. In January of 1919 a copied A7V tank turned up in Berlin. It had no cannon, but rather a 08/15 machine gun at each corner in a casemate-type mantlet. There were two doors on each side, and the turret was very different. The tank was used on January 15, 1919 when the loyal troops marched into Berlin (upper and lower left photos), and later appeared in Leipzig under the name "Hedi."

Of the 100 ordered A7V chassis, 22 were used for the A7V tanks and another (No. 524) served as the chassis of the A7V-U. The others were built as cross-country trucks.

Number 514 (Army Mobile Column — Caterpillar — 1111) demonstrates its cross-country mobility.

Below: High gasoline consumption and heavy wear limited the utility of these cross-country vehicles very much. Their loads often slid too, which led to damage.

Below: Orenstein & Koppel of Berlin and Weserhütte of Bad Oeynhausen developed a ditch-digger on a modified A7V chassis. Sixty to eighty of them were built.

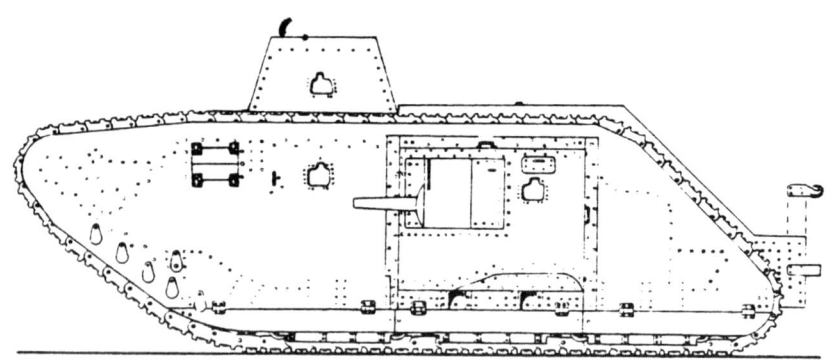

Above: Drawing of the A7V-U. The commander and driver were now located behind the front plate of the hull. In the turret (with six gun ports) was a machine gunner's seat.

Right: Rear view of the A7V-U with its two side bays, with gun ports set into the entrance doors. The number 524 is that of the chassis which had been rebuilt for this tank. The first driving tests of this vehicle, which weighed almost 40 tons, took place on June 25, 1918.

Below: The A7V-U, seen at an angle from the front, shows the construction of one of the bays with a Belgian 5.7-cm cannon. Overall the tank was longer (8.38 meters) and wider (4.69 meters) than the A7V. On September 12, 1918 the OHL ordered work stopped and the vehicle dismantled.

Because of the favorable handling qualities of the British tanks, the OHL ordered in March of 1917 that an A7V should be built with cyclical tracks. The one example that was built turned out to be too big and heavy and was inclined to tip forward when off the road. The rhombus form was not retained.

The first driving tests on May 25, 1918 showed the weakness also seen in British tanks: the drive track in the upper part of the tank was soon fouled with sand. In addition, the driving opposition was 40% higher than in the basic version. On September 12, 1918 this project was abandoned.

LARGE TANK (K-WAGEN)

Even before the first A7V was finished, the Chief of Motor Vehicles (Chefkraft) of the War Ministry contracted for the construction of the so-called K-Wagen.

This project, which can be described as completely nonsensical from the start, tied up not only production capacities but also raw materials in the ensuing time. At that time too, the sufficiently knowledgeable tactician must have wondered how a vehicle 13 meters long, weighing well over 100 tons, could be put to use on a battlefield torn up by artillery fire.

In a technical sense at least, this tank deserves interest. The construction of ten K-Wagen was approved by the War Ministry on June 28, 1917. The requirements included a trench-spanning capability of four meters, an armor up to three centimeters thick (on the front and sides), an armament of one or two semi-automatic cannons (5- to 7-cm caliber), four machine guns, two flame-throwers, and an 18-man crew. The overly heavy vehicle was to be driven by 400 HP (!) — two motors producing 200 HP each. The vehicle was to be transported in single loads of about 30 tons and assembled just a few kilometers behind the front.

The mass of criticism that came forth at the first commission discussions was disregarded. The suggested construction time of one year was cut to eight months at the request of the OHL.

At the very beginning of designing, a multitude of problems appeared, as no previously used components or building techniques could be utilized. Thus the vehicle was to be built by bridgebuilding companies. Only machine toolmakers could produce the gears. The clutches also had to be made completely new. The tracks derived from power shovel construction.

The hopeless notion of attaining sufficient motive power with 400 HP had to be corrected quickly. Finally two Daimler 6-cylinder marine engines producing 650 HP each were chosen.

Meanwhile the weight had risen to 140 to 150 tons. Shortening the vehicle gave a planned weight of 120 tons.

Four 7.7-cm cannons from Idstein Fortress were originally made available. Contracts to build five vehicles each were given to the Riebe Ball Bearing Works in Berlin-Weissensee and the Waggonfabrik Wegmann in Kassel.

In view of the first experiences in driving technology, such as with the A7V, and the successful Allied use of tanks on the western front, the government offices seem to have developed serious doubts as to the practicality of the K-Wagen. The test department of the field vehicle office stated on October 18, 1917 that the K-Wagen was suitable only for positional warfare.

The program suffered constant delays, mainly on account of technical problems and difficulties with suppliers. When the war ended, one vehicle was almost finished at the Riebe works, a second (without motors) less far along, while at the Wegmann works one armored body was nearly finished.

The drawing shows the division of space in the K-Wagen; the two drivers are located at the front with three machine guns; in the fighting compartment behind them are the four 7.7-cm cannons mounted in the side bays, plus two more machine guns; in the cylindrical cupola on the roof are the commander and an artillery officer; in front of the engine compartment is a signalman; behind him is the engine compartment with two machinists, each with a machine-gun bay to the rear; farthest to the rear is the gearbox.

This side view with the side bays removed affords a view into the interior.

The K-Wagen, seen from the right rear, shows the two exhaust mufflers with the air intake openings behind them.

Below: This rear view makes clear the large-surface plates of the track links.

Below: This front view shows the low design of the cupola, which caused a large dead angle for the commander.

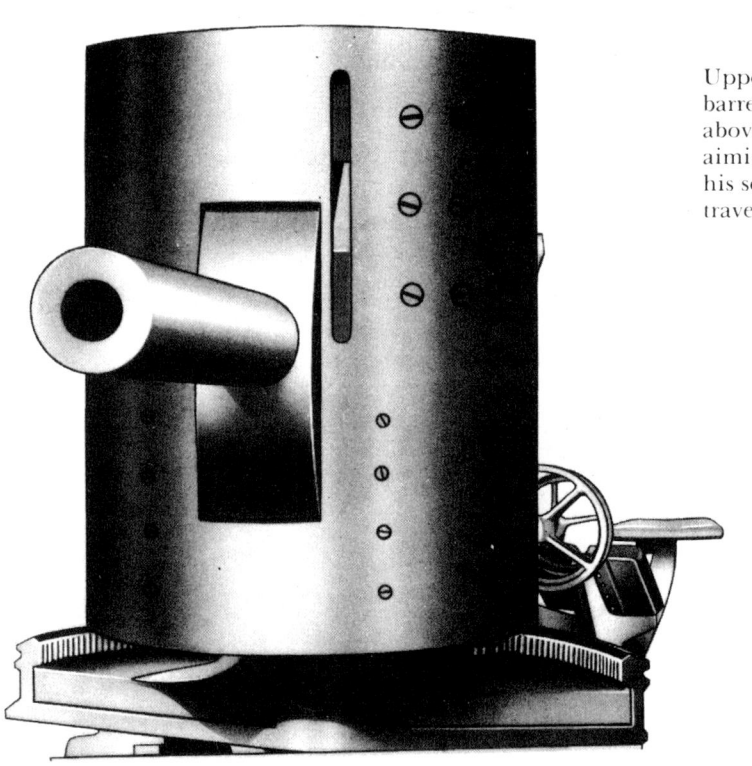

Upper right: The socket mantlet with built-in barrel and adapted cylindrical shield. At the left, above the cam lock that opens to the left, the aiming scope can be seen. The gunner, sitting on his seat attached to the mantlet, goes through all traverse aiming movements.

Above: Front view of the tank gun. The vertical long slit affords a clear view of the target area for the aiming scope, which follows the aiming motions.

Right: The mantlet with the shield bearings at the top. The handwheel controls traverse aiming.

Below: The 7.7-cm casemate cannon when removed. Under the barrel is the combined recuperator and barrel brake, under it the toothed wheel of the elevation aiming mechanism. The right shield bearing can be seen on the side.

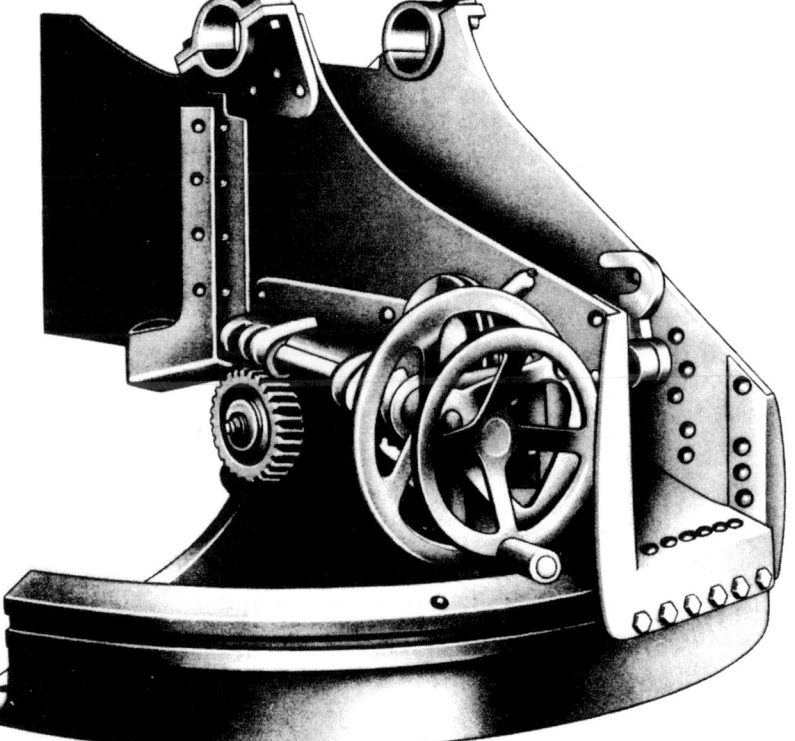

LIGHT TANK

Under the impression that light French and (later also) British tanks were being used successfully, Chefkraft suggested on December 29, 1917 that light tanks be built by using motor vehicle chassis, of which many were available. This was rejected by the OHL on January 17, 1918 because of too-light armor.

Meanwhile the OHL worked with Krupp on its own light tank project — without the cooperation of Chefkraft, whose often limited competence in the field had inspired comment from all sides. When Chefkraft finally had to be involved, it turned out that this office had promoted the development on its own.

The vehicle, weighing about eight tons, was to be able to reach 12 to 15 kph and climb grades of about 45 degrees. Trenches up to two meters wide should be crossed. The armament of this tank, designated "Leichter Kampfwagen I", was to be a 5.7-cm rapid-firing cannon.

Test drives with the first prototype in March of 1918 went off satisfactorily; the designing of an improved second model — LK II — was begun; it was to have, among other things, heavier armor. Its weight of eight tons was two tons greater.

On June 13, 1918 both models, the LK and the Krupp-Wagen, were presented. Both tanks were to be developed further until ready for series production, at first only as machine-gun cars. The LK was to be produced later with a cannon too.

On July 23, 1918 Krupp presented the specifications of a new small assault tank that was bigger than the former design and was to be armed with a 5.2-cm cannon and a machine gun.

With tracks 1.1 (later 1.3) meters longer, the LK II was considerably better for cross-country use than the Krupp design (1.5-meter tracks), so that after the completion of one test model on October 2, 1918, the order for the Krupp model (65 units) was cancelled.

The beginning of series production of the LK II (with machine guns) was planned for December 1918, with the production planned

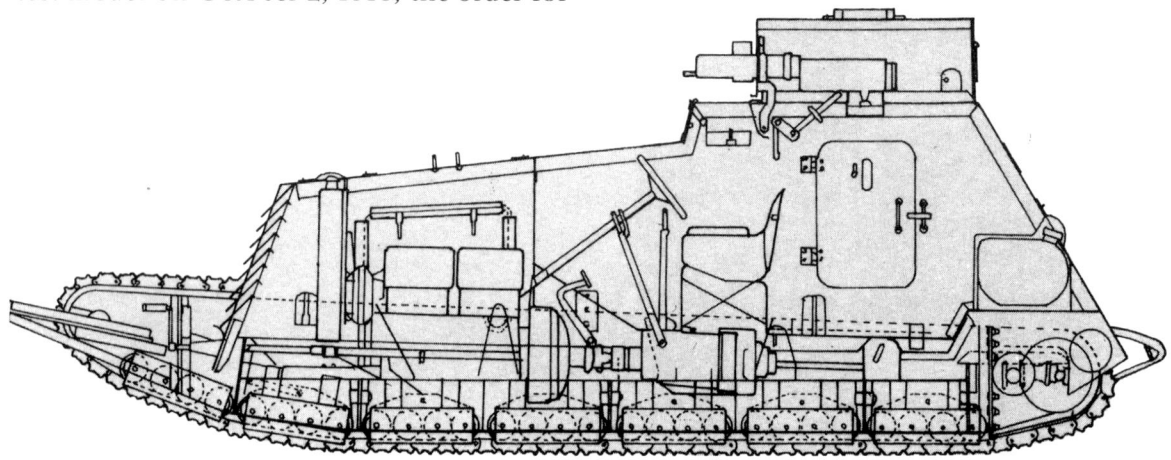

This see-through drawing shows the division of space, with the front engine and the rotating turret at the rear.

Below: A side view of the LK 1 with its entry doors and engine servicing panels open. The soldier standing at the rear gives an impression of its compact construction.

for ten units. After that, production was to increase steadily, with 200 units produced per months beginning in April of 1919. On August 29, 1918, after firing tests with an **LK II** prototype (armed with a 5.7-cm cannon), it was announced that the body was too weak. Instead of it, the OHL ordered on September 30, 1918, that the Krupp 3.7-cm KwK be installed, with only one-third of the tanks to be equipped with machine guns.

By the war's end a few vehicles were finished, a great many under construction. Several **LK II**'s were sold to Sweden, where they were modified as "Stridswagen m21 light" and used for years. Shortly before the war ended, Vollmer presented a design for the **LK III**, which was to have the motor and gearbox located in the rear.

LK I traveling cross-country.

Lower left: Front view of the LK I with the radiator grille at the front, closed by seven transverse ribs of armor plate.

Below: An angled front five of the LK I. The two-piece driver's flap is closed. With only five transverse cutouts, the driver had only a modest field of vision forward.

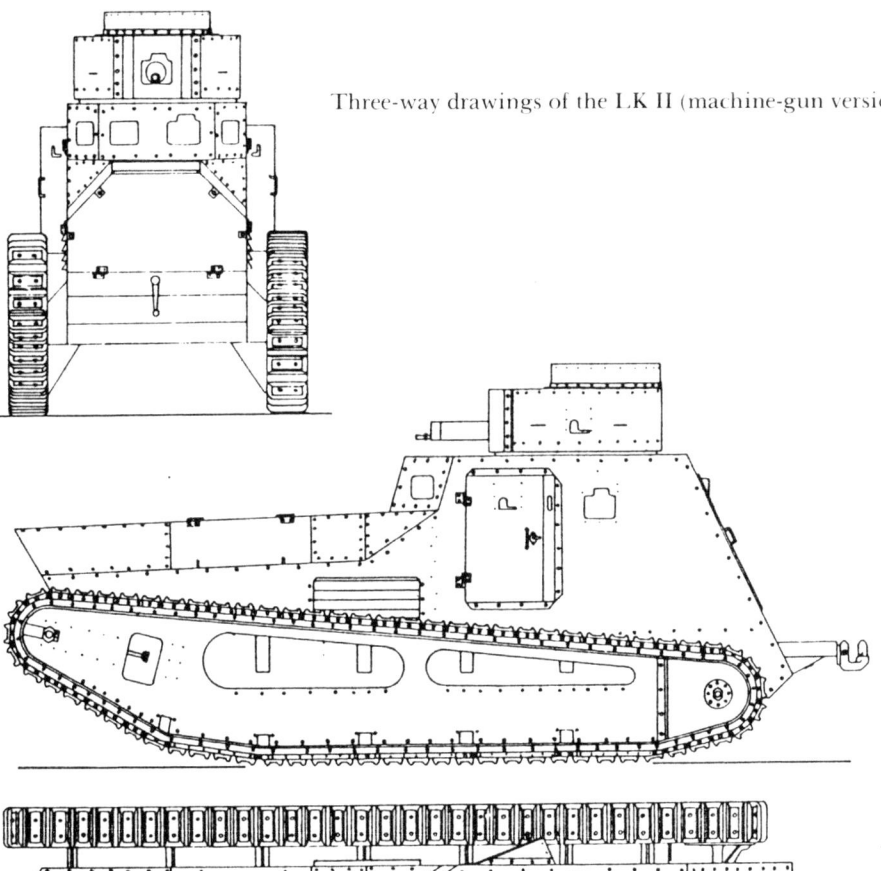

Three-way drawings of the LK II (machine-gun version).

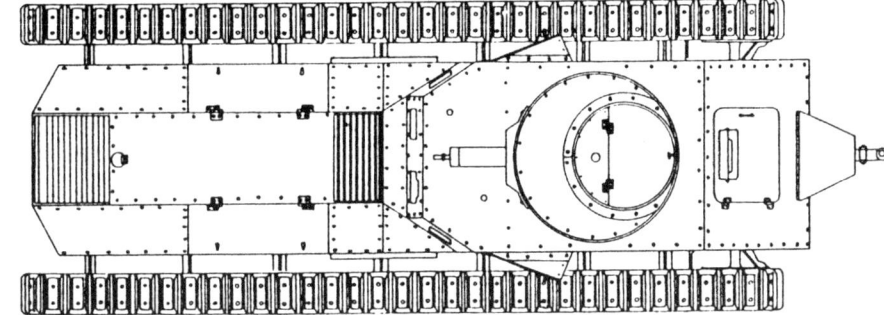

Upper left: This rare photo shows an LK II on a test run at Marienfelde in October of 1918. The armored body is still lacking. At the stern is the gearbox with the driveshafts running along both sides, the gears at their ends acting directly on the (larger upper) drive wheels at the end of the running gear. The two lever arms at the rear control the drive brakes (left) and the setting brakes (longer right lever).

Left: This LK II model shows the entire body of the machine-gun tank.

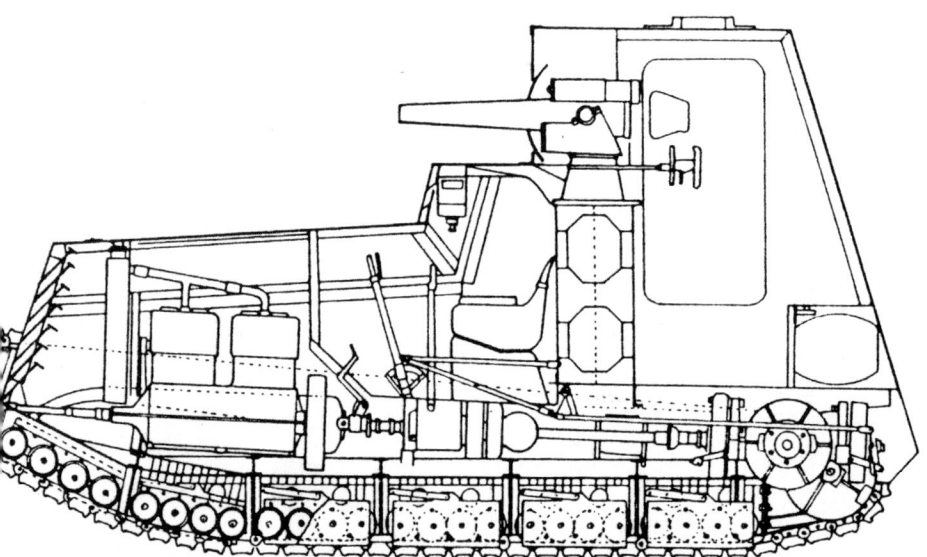

Above: Drawing of the LK II cannon version.

LK II (Kanonenversion)

Lower left: The cannon version of LK II seen from the right. One machine gun can be used from each of the side doors.

Below: The same tank from the front. The 3.7-cm cannon is partially movable. The arc of traverse is 30 degrees to either side.

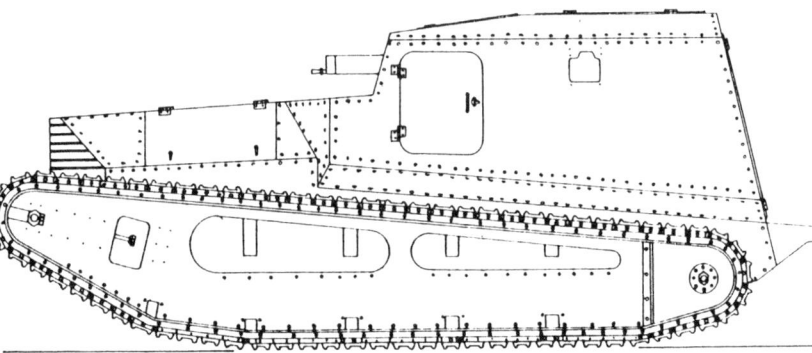

Above: To get around the armistice stipulations, ten LK II tanks were finished and delivered to Sweden in the autumn of 1921. There they were used for some years, known as Stridsvagn m21. One last example is in the Axvall Tank Museum.

Left: Light towing tractors or powered limbers for artillery guns were suggested by both Krupp and Vollmer. These were armed with one machine gun pointing forward. The drawing shows the LK II power limber, Vollmer type. There was barely room for a six-man crew in the rear; the driver sat at the right front, with the machine gunner beside him.

ASSAULT TANK "OBERSCHLESIEN"

The technically unsatisfactory conception of the A7V led in 1918 to a series of further tank projects, of which the "Oberschlesien" (Upper Silesia) assault tank was especially interesting and very progressive.

Chefkraft too had been convinced by June of 1918 that building tanks of the heaviest type had to be abolished to concentrate on light, mobile assault tanks (cannon plus machine guns) that could be mass-produced. Despite that, the ten ordered K-Wagen tanks were not cancelled, though they used up immense quantities of raw materials.

Besides these two types, two others were considered, namely heavy assault tanks (such as the A7V or captured tanks) and armored power limbers of the lightest type, armed with machine guns, to move infantry guns or mine throwers.

Thirteen firms competed to build a heavy assault tank, but scarcely one project got beyond the preliminary stage. The firm of Louis Ehlers designed the "Hannover" tank, while the Wegmann firm planned the "Hessen-Cassel" infantry tank, both presented in October of 1918.

A member of Vakraft (Hauptmann Müller) offered the design of a tank, in the middle of 1918, that wastaken over by the Oberschlesische Hüttenwerke (Upper Silesian Steel Works) in Gleiwitz. It was given the disguise name of "Oberschlesien." The drawings of this tank show a well-conceived overall design that was to become the classic tank design in the years to come: Driver up front, separated fighting and rear engine compartments, raised central weapons in a rotating turret.

The ground pressure of the tank was only about 0.5 kg/cc with a weight of some 20 tons. Great care was given to the running gear, the drive wheels of which were located in the middle of the tank. At both ends of the running gear were geared arcs that encircled a fixed drum.

The power source was an Argus aircraft motor (Type AS 3), which produced 180 HP at 1400 rpm. The fuel tanks were now located outside the crew's area. On October 5, 1918 the OHL determined that two test models were to be built at first, but this did not take place.

The production of two modified prototypes, "Oberschlesien II" with caterpillar tracks and rear drive wheels, suggested by Vakraft on Octoberr 12, 1918, did not take place either.

Precise information on other heavy tanks, such as the "Horch" or "Benz-Bräuner", is lacking.

Assault tank "Oberschlesien"

TECHNICAL DATA	A7V	K-Wagen	LK II/ cannon	Oberschleisen II
Weight (tons)	30	150	8.5	19
Length (mm)	7350	12700	5080	6700
Width (mm)	3050	6000	1950	2340
Height (mm)	3300	3000	2675	2965
Clearance (mm)	200		400	600
Engine	Daimler	Daimler	Various	Argus
Horsepower	2x100	2x650	40-60	180
Speed (kph)	10	7.5	14	16
Fuel (liters)	500	3000	150	1000
Range (km)	35	no data	70	no data
Crew	18	22	3	5
Armament				
Cannon caliber	5.7	4x7.7	3.7	5.7
Rounds	180	8000	100	no data
Machine guns	6	7	0	2

„UHU"-He 219

THE BEST NIGHT FIGHTER
OF WORLD WAR II

SCHIFFER MILITARY

ARMORED MILITARY
VEHICLES

MAUS

AND OTHER GERMAN ARMORED PROJECTS

DORNIER DO 335
"PFEIL"

THE LAST AND BEST PISTON-ENGINE
FIGHTER OF THE LUFTWAFFE

GERMAN

ARMORED TRAINS

IN WORLD WAR II

GERMAN MOTORCYCLES
IN WORLD WAR II

SCHIFFER MILITARY

ALSO FROM:

•SCHIFFER MILITARY HISTORY•

•THE WAFFEN-SS•THE HG PANZER DIVISION•
•THE 1ST SS ARMORED DIVISION•
•THE 12TH SS ARMORED DIVISION•

AND MORE...

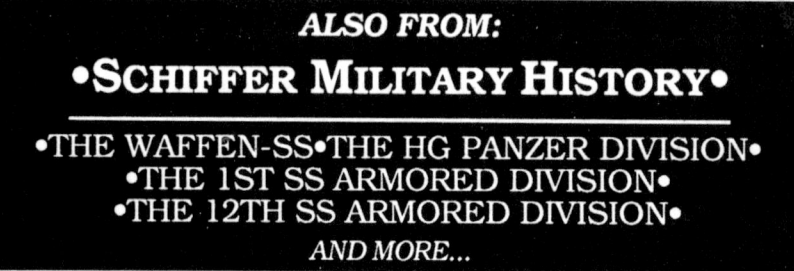

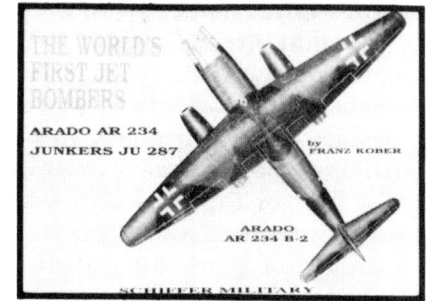

THE WORLD'S
FIRST JET
BOMBERS

ARADO AR 234
JUNKERS JU 287

by
FRANZ KOBER

ARADO
AR 234 B-2

SCHIFFER MILITARY

GERMAN AIRSHIPS

PARSEVAL-SCHUTTE-LANZ-ZEPPELIN

HEINZ J. NOWARRA   -SCHIFFER MILITARY-

GERMAN BATTLETANKS

"NEWLY BUILT VEHICLE"-PANZER I-PANZER II-PANZER III
-PANZER IV- PANZER V "PANTHER"-PANZER VI "TIGER"
and "KING TIGER"-"MAUS"

IN COLOR
1934-45

GERMAN
NIGHT
FIGHTERS
IN WORLD WAR II

MANFRED GRIEHL

Ar 234-Do 217-Do 335-Ta 154-He 219-Ju 88-Ju388-Bf 110-Me 262 etc.

-SCHIFFER MILITARY-

The King Tiger Tank

by Horst Scheibert

THE PANTHER FAMILY

by
HORST SCHEIBERT

SCHIFFER MILITARY

GERMAN
PERSONNEL CARS
WARTIME

The best of
medium and heavy
personnel vehicles
of the era

CAPTURED TANKS UNDER
THE GERMAN FLAG

RUSSIAN BATTLE TANKS

HORST SCHEIBERT

-SCHIFFER MILITARY-

GERMAN SHORT-RANGE
RECONNAISSANCE PLANES
1930-1945